ISBN: 978-1647904098(pb); 978-1647904807 (hc)
Publisher: Dylanna Publishing, Inc.
First Edition: 2024

10 9 8 7 6 5 4 3 2 1

For information about special discounts for bulk purchases, please contact:
orders@dylannapublishing.com
Dylanna Publishing, Inc.
www.dylannapublishing.com

DYLANNA
Press

Arctic Animals

Meet the Amazing Creatures of the Polar North

Tyler Grady

Contents

Welcome, young explorers, to one of the coolest places on Earth—literally! Prepare to explore the Arctic, a remarkable region located at the northernmost part of our planet. This unique environment is characterized by its perennially frozen ground and its icy seas, which are dotted with colossal icebergs of various shapes and sizes. The Arctic is a place of extreme conditions, where temperatures can plummet to staggering lows and the sun may not rise for months during the winter. Despite these harsh circumstances, the Arctic is home to a diverse array of fascinating wildlife that has adapted to thrive in this challenging ecosystem.

The Arctic isn't just a single spot, but a vast area that stretches over parts of eight countries—Canada, Greenland, Iceland, Norway, Sweden, Finland, Russia, and the United States (thanks to Alaska). This wide ring of frosty lands and icy waters is home to some of the most incredible landscapes and creatures you could ever imagine.

But the Arctic is more than just ice. It's a land of contrasts. In the summer, parts of it bloom with wildflowers, and the sun never sets, bathing everything in a never-ending golden light. This phenomenon is called the Midnight Sun, where for several weeks, the sun barely dips below the horizon, if at all. Then, in the winter, the same places plunge into deep darkness during the Polar Night, when the sun doesn't rise for months.

The Arctic Ocean is the heart of this region, surrounded by a fringe of land covered in **permafrost.** Permafrost is ground that remains completely frozen for two or more years straight. Some of it has been frozen for thousands of years! This icy environment creates a unique ground for scientists to study, offering clues about the Earth's ancient climates and even the possibility of life on other planets.

Despite the cold, the Arctic is teeming with life. It's a place where you can find polar bears, the largest land predators on Earth, wandering across the ice in search of their next meal. The waters below are bustling with whales, seals, and walruses, while the skies are crossed by migratory birds traveling thousands of miles to breed and feed in the Arctic summer.

Map of Arctic region.

This icy wonderland is not only a place of stunning natural beauty and fascinating animals but also an important area for scientists studying climate change. The changes happening in the Arctic affect the whole planet, making our journey here even more crucial. So zip up your warmest coat, and let's dive into the wonders of the Arctic together!

What Makes the Arctic Unique?

The Arctic is like no other place on Earth, and its uniqueness comes from both its environment and the life that thrives there. Here's why this chilly corner of our planet stands out:

DID YOU KNOW?

The Arctic gets its name from the Greek word "arktos," meaning bear. This isn't just any bear, but the constellation Ursa Major, or the Great Bear, which points to the North Star, guiding explorers to the Arctic!

Extreme Day and Night Cycles: The Arctic experiences dramatic changes in daylight throughout the year. During the summer months, the sun doesn't set for days or even months at a time, creating what's known as the Midnight Sun—a period where the sun is visible at midnight. Then, in winter, the opposite happens. The sun disappears for a long stretch, leaving the Arctic in darkness, known as the Polar Night. This cycle affects everything from the behavior of animals to the lives of people in the Arctic.

Climate and Temperature: Even though it's known for its cold, the Arctic's climate can vary widely. In winter, temperatures can plummet to below -50°C (-58°F) in some areas. However, during the short summer, parts of

Alaskan mountains in summer.

Arctic landscape.

the Arctic can warm up to 10°C (50°F), allowing a burst of life to flourish.

Ice, Ice, and More Ice: The Arctic is home to some of the most spectacular ice formations on the planet, including sea ice that covers the ocean surface, glaciers that carve the landscape, and permafrost. Sea ice is particularly important because it acts as a platform for animals like seals and polar bears and helps regulate the Earth's temperature by reflecting sunlight back into space.

Unique Ecosystems: The Arctic isn't a barren wasteland of ice; it's home to diverse ecosystems. From the tundra, where hardy plants and flowers bloom in the brief summer, to the Arctic Ocean, teeming with marine life like fish, seals, and whales, the Arctic supports a surprising variety of life. Each species has adapted in fascinating ways to survive in this extreme environment.

Environmental Indicator: The Arctic is on the front lines of climate change, warming at twice the rate of the rest of the planet. This has profound implications for the rest of the world, affecting global weather patterns and sea levels.

Understanding what makes the Arctic unique helps us appreciate its importance to the global ecosystem and the urgency of protecting this fragile environment for future generations.

DID YOU KNOW?

In summer, the Arctic can have 24 hours of sunlight, called the "Midnight Sun." In winter, it can be dark all day long, known as the "Polar Night."

Northern Lights (aurora borealis)

Our Arctic Adventure Begins

Our journey to the Arctic isn't just about seeing a world covered in ice and snow or watching the sky light up with the Northern Lights. We're here to meet the remarkable animals that thrive in this icy realm. Each of these animals, from the mighty polar bear to the quick-witted Arctic fox, has incredible adaptations that allow them to live, play, and survive in one of the planet's coldest environments.

Some of the animals we'll be covering include:

Polar Bears: Known as the "kings of the Arctic," polar bears are the largest land carnivores. They have a thick layer of fat and a waterproof coat that keeps them warm in freezing temperatures. Their large paws help them to tread on thin ice and swim long distances between ice floes to hunt seals.

Arctic Foxes: These small, resourceful foxes change color with the seasons. In winter, they wear a white coat that blends into the snow, hiding them from predators and prey. In summer, they turn a brown or gray color to match the tundra's rocks and plants. Arctic foxes have furry soles to keep their paws warm on the icy ground.

Walruses: With their distinctive long tusks and whiskered faces, walruses are one of the Arctic's most iconic animals. They use their tusks to pull themselves out of the water and onto the ice. Walruses are social animals, often found huddled together in large groups to keep warm.

Snowy Owls: These magnificent birds are well adapted to the Arctic life, with their white feathers providing camouflage against the snow. Snowy owls are among the few owl species that are **diurnal,** meaning they hunt both at night and during the day, depending on the availability of prey.

Narwhals: Often called the "unicorns of the sea," narwhals are known for the long, spiral tusk that protrudes from their heads. This tusk is actually a tooth, and scientists believe it may be used to sense changes in the environment, helping narwhals navigate the icy Arctic waters.

Caribou: Also known as reindeer in some parts of the world, caribou undertake one of the longest land migrations of any animal, traveling hundreds of miles between their wintering and summering grounds. Their large hooves act like snowshoes, supporting them on soft snow and wet tundra.

This Arctic adventure will introduce us to these incredible animals and more, revealing how they've mastered the art of survival in the Arctic's extreme conditions. Let's dive into their world and uncover the secrets of their resilience and ingenuity.

Polar Bear: The Arctic's Majestic Hunter

In the vast, icy world of the Arctic, the polar bear reigns as the largest land predator, a true marvel of nature's design. Imagine a creature so perfectly adapted to its freezing environment that it can disappear into the landscape, thanks to its seemingly white fur that's not really white at all! Each strand of fur is a clear, hollow tube that reflects light, **camouflaging** the bear against the snow and ice. Underneath this remarkable coat lies a thick layer of blubber, insulating the bear against the bone-chilling temperatures of its home.

A polar bear's life is one of solitude, wandering the vast expanses of ice-covered oceans in search of food. Their diet is mainly seals, which they hunt with incredible patience and skill. They wait silently at a seal's breathing hole, ready to strike with their powerful paws, or they may even swim underwater to surprise their prey. Speaking of swimming, polar bears are exceptional at it, using their large, webbed front paws to propel themselves through the water, covering distances that can span hundreds of miles.

The Arctic winter is a time of birth for polar bears. After mating in the spring, a female polar bear will give birth to one to three cubs in the deep snow dens of the Arctic tundra. These cubs, tiny and vulnerable at birth, stay with their mother for about two and a half years. During this time, she teaches them everything they need to know to survive in one of the harshest environments on Earth.

Communication among polar bears is a fascinating aspect of their behavior. Though they prefer the life of a loner, they do communicate through various sounds like growls, chuffs, and roars, especially during the mating season or

Fast Facts!

- **Master of Camouflage:** Despite looking like they're dressed in a snowy white coat, polar bear fur is actually clear! It's the fur's special structure that scatters and reflects light, making them appear white and perfectly camouflaged in their icy home.
- **Gigantic Swimmers:** Imagine swimming across an entire city! Polar bears can swim hundreds of miles without stopping, using their large, webbed paws like natural paddles to glide through the icy waters.
- **The Arctic's Top Predator:** With their massive size, males can weigh as much as a small car, and their sharp claws and strong teeth make them the ultimate predator on the ice.
- **Sniffing Out Dinner:** A polar bear's nose is so powerful, it can smell a seal's den buried under 3 feet of snow and ice from a mile away. Talk about a super sniffer!
- **Nature's Snowshoe:** Their huge paws aren't just for swimming; they also act like snowshoes, spreading out their weight so they don't sink into the snow. This makes them excellent at tiptoeing across thin ice.
- **Built-in Winter Coat:** Polar bears have a thick layer of fat, called blubber, that keeps them warm in the freezing Arctic temperatures. It's like wearing a snug, insulating jacket all year round.
- **Long-Distance Migrants:** Some polar bears travel over 600 miles in search of food, making them one of the animal kingdom's great wanderers. That's like walking from New York City to Detroit!
- **Solar-Powered Bears?**: Not really, but their black skin absorbs the sun's rays, helping them stay warm. The sun's energy is trapped by their skin, beneath all that fur and blubber.
- **Underwater Hunters:** Polar bears can hold their breath for up to two minutes while diving underwater. This skill comes in handy when they're hunting for seals, their favorite meal.
- **Swim between Ice Floes:** Polar bears use ice floes as platforms to rest, hunt, and sometimes give birth, showcasing their dependence on sea ice for survival.

All About Polar Bears

Scientific Name: *Ursus maritimus*

Type: Marine Mammal

Size: Largest bears in the world, with males weighing between 800 to 1200 pounds (360 to 550 kg) and females weighing from 300 to 700 pounds (135 to 320 kg). They stand about 5 feet (1.5 meters) tall to the shoulder and range in length from 6 to more than 9 feet (1.8 to 2.75 meters).

Diet: Carnivores, feed mostly only seals. When food is scarce they will eat anything they can find, including fish, rodents, birds, eggs, and even reindeer.

Breeding and Young:
Typically reproduce every 3 years. Females give birth to 1-3 cubs. Cubs stay with their mother for about two and a half years, learning how to hunt and survive in the Arctic.

Social Life and Communication: Polar bears are mostly solitary animals, except for mothers with cubs or during mating season. They communicate through vocalizations like growls, chuffs, and roars, and through scent marks left in their tracks.

Territory: Nomadic, travel long distances to find food

Sleep: 7-8 hours per day

Habitat: They primarily inhabit the ice-covered waters of the Arctic Ocean and its surrounding seas and coastlines.

Lifespan: 15-18 years in the wild

Status: Vulnerable, population is in decline

when mothers interact with their cubs. They also use scent marks to leave messages for other bears, perhaps as a way to avoid unwanted encounters.

Polar bears don't have a permanent home and can often be found napping in the snow after a successful hunt, using the thick blanket of snow as insulation against the cold. Their **territories** are vast, sometimes spanning hundreds of thousands of square kilometers, as they follow the seasonal movements of the sea ice to hunt for seals.

One of the most remarkable facts about polar bears is their ability to fast for months. During the summer and fall, when the sea ice melts away, and hunting becomes more challenging, they rely on the fat reserves built up during the hunting season. This incredible adaptation ensures their survival in a landscape that is rapidly changing due to climate change, a threat that endangers their very existence by reducing their hunting grounds.

DID YOU KNOW?

Polar bears can smell their prey up to a mile away and through several feet of solid ice. This extraordinary sense of smell is crucial for their survival in the vast, open landscapes of the Arctic.

Polar bears, with their majestic presence and incredible adaptations, are not just survivors; they are the kings and queens of the Arctic ice. But as their world changes, they remind us of the importance of taking care of our planet, ensuring that the icy realms of the Arctic continue to thrive for generations to come.

Arctic Fox: The Arctic's Enchanting Survivor

The Arctic fox is a small but amazing animal, built to survive in one of the toughest places on Earth. Imagine a creature that changes color with the seasons—white in winter to blend in with the snow, and brown or gray in summer to match the rocky tundra. This clever camouflage helps the Arctic fox stay hidden from predators and sneak up on its prey.

Arctic foxes are small, weighing between 6 to 10 pounds (2.7 to 4.5 kg) and measuring about 20 to 30 inches (50 to 76 cm) long, not counting their bushy tails. Their round, compact bodies help them keep warm, and their short ears and noses reduce heat loss. They even have furry soles on their feet to prevent slipping on ice and to keep warm.

These foxes are not picky eaters. In winter, they often follow polar bears to scavenge leftovers from their kills. In other seasons, they hunt rodents like lemmings, birds, and even fish. They also eat berries and seaweed when they can find them.

One of the most important adaptations of the Arctic fox is its thick fur, which keeps it warm even in temperatures as low as -58 degrees Fahrenheit (-50 degrees Celsius). Their fur-covered feet act like built-in snow boots, providing warmth and traction on icy surfaces.

Arctic foxes are loyal partners, mating for life. They return to the same den year after year to raise their babies, which are called kits. A mother fox typically gives birth to 5-10 kits at a time. Both parents will help raise the young. The mother fox will stay in the den with the baby foxes while the father will go out to hunt for food.

Fast Facts!

- **Color-Changing Fur:** Arctic foxes change their fur color with the seasons—white in winter to blend with the snow, and brown or gray in summer to match the tundra.
- **Furry Feet:** Their feet are covered in fur, acting like built-in snow boots for warmth and traction on ice.
- **Omnivorous Diet:** They eat a variety of foods including rodents, birds, fish, berries, and even seaweed, and often scavenge leftovers from polar bear kills in winter.
- **Active Year-Round:** Arctic foxes do not hibernate and are active throughout the year, even in the harsh Arctic winter.
- **Ecosystem Engineers:** By storing extra food in the summer, they help fertilize the soil and support plant growth.
- **Incredible Insulation:** Their thick fur and fat layer keep them warm in temperatures as low as -58 degrees Fahrenheit (-50 degrees Celsius).
- **Fast Runners:** Despite their small size, Arctic foxes can run up to 30 miles per hour (48 kilometers per hour) when they need to.
- **Nomadic Lifestyle:** Some Arctic foxes travel great distances, even hundreds of miles, following food sources and changing conditions.
- **Sharp Hearing:** Their sharp hearing allows them to locate prey under the snow, which they catch by diving in a behavior known as "mousing."
- **Seasonal Fur Shedding:** They shed their thick winter fur in the spring, replacing it with a shorter, lighter summer coat.
- **Super Smellers:** Their sense of smell is so keen that they can locate carcasses under deep snow, which helps them find food in winter.
- **Solitary Hunters:** Although they are social within family groups, Arctic foxes usually hunt alone, using their keen senses to track down prey.
- **Winter Survival:** During the harshest winter months, Arctic foxes can reduce their metabolic rate to conserve energy, allowing them to survive longer without food.
- **Resilient Travelers:** Arctic foxes have been known to travel across sea ice and even swim between ice floes to reach new hunting grounds.

All About Arctic Foxes

Scientific Name: *Vulpes lagopus*

Type: Mammal

Size: Arctic foxes are small animals, roughly the size of a housecat, with a body length of 20 to 30 inches (50 to 76 cm) and a height of about 12 inches (30 cm) at the shoulder. They weigh between 6 to 10 pounds (2.7 to 4.5 kg).

Diet: Omnivores, primarily feeding on small mammals like lemmings. They also eat birds, eggs, insects, and berries. In winter, they scavenge on polar bear kills.

Breeding and Young:
Arctic foxes breed once a year. Females give birth to a large litter of 5 to 8 pups. Pups are weaned at 6 weeks and start hunting at 3 months.

Social Life and Communication: Arctic foxes have complex social structures during breeding season, living in underground burrows with extended family members. They communicate with a series of barks, howls, and whines.

Territory: Large territories that they mark with scent markings, but they can be nomadic in search of food, particularly in winter.

Sleep: Varies greatly, depending on the season; they may be more active during the night in the summer months

Habitat: Found throughout the Arctic tundra, from the edge of the forest into the deep polar regions.

Lifespan: 3-6 years in the wild

Status: Not endangered

Unlike other types of foxes, Arctic foxes are social animals. They make different sounds to communicate, especially during mating season or when defending their territory. A group of foxes is called a skulk.

DID YOU KNOW?

When food is plentiful, Arctic foxes will bury excess food to store for later. This helps them survive during times when food is hard to find.

Arctic foxes don't hibernate; they're active all year long. They roam large territories in search of food. When it's time to sleep, they curl up in a ball with their fluffy tails wrapped around them for warmth, often resting in the open. Sometimes they will dig tunnels in the deep snow for shelter.

Arctic foxes live all over the Arctic tundra, from Scandinavia and Russia to Greenland and Canada. They're adaptable and can live on cliffs, open tundra, and even sea ice.

Arctic foxes show us how to survive and thrive in the harshest conditions. They're small, but they're tough and incredibly resourceful, making them one of the Arctic's most impressive inhabitants.

Walrus: The Arctic's Gentle Giant

In the chilly waters of the Arctic, the walrus stands out with its massive body, long tusks, and bristly mustache. These fascinating creatures are well adapted to life in some of the coldest and harshest environments on Earth.

Walruses are enormous—weighing up to 3700 pounds (1700 kg) and growing to 12 feet (3.5 meters) long. Their most iconic feature is their long tusks, which can reach up to 3 feet (1 meter) in length. These tusks, which are actually elongated canine teeth, are used for a variety of purposes, including helping them haul their massive bodies out of the water, breaking through ice, and even for defense against predators like polar bears and orcas.

A walrus's diet is quite varied, but they mainly feed on clams, snails, worms, and other small marine animals found on the ocean floor. Using their sensitive whiskers, called vibrissae, they can detect prey hidden in the sediment. They then use their powerful suction to pull the soft-bodied animals out of their shells, often swallowing their meals whole.

Walruses are built for the cold. Their thick blubber not only insulates them against the freezing temperatures but also helps them stay buoyant in the water. Their skin, which can be up to 4 inches (10 centimeters) thick, adds another layer of protection against the cold.

Breeding season for walruses occurs in the winter, with males using their tusks and **vocalizations** to compete for females. After a gestation period of about 15 to 16 months, the female gives birth to a single calf, which she will nurse for up to two years. Walrus calves are born with a thin layer of blubber and can swim almost immediately, though they stay close to their mothers for warmth and protection.

Walruses are highly social animals, often seen in large groups called herds. These herds can range from just a few individuals to several thousand. Within these groups, walruses communicate using a variety of sounds, including barks, grunts, whistles, and bell-like calls. Males are particularly vocal during the breeding season, using their distinctive sounds to attract females and establish dominance over other males.

Fast Facts!

- **Massive Marine Animals:** Walruses are among the largest marine mammals in the Arctic, with adult males weighing up to 3,700 pounds and growing up to 12 feet long.
- **Iconic Tusks:** Their tusks, which can grow up to 3 feet long, are used for combat, dominance displays, and helping them haul out onto ice or land.
- **Sensitive Whiskers:** Walruses have whiskers (vibrissae) that are highly sensitive and help them detect and find shellfish on the ocean floor.
- **Deep Divers:** They can dive to depths of over 260 feet (80 meters) in search of food, holding their breath for up to 30 minutes.
- **Clam Connoisseurs:** Their diet mainly consists of bottom-dwelling sea creatures, especially clams, which they suck out of the shell using powerful lips.
- **Social Creatures:** Walruses are very social and often gather in large herds of hundreds or even thousands on ice or beaches.
- **Arctic and Subarctic Residents:** They primarily live in the cold waters of the Arctic Ocean and the northern seas off Russia, Canada, Greenland, and Alaska.
- **Thick Skin and Blubber**: They have a thick layer of blubber (up to 6 inches) under their skin, which keeps them warm in icy waters and serves as an energy reserve.
- **Cultural Significance:** Walruses have been of great importance to Indigenous peoples of the Arctic for thousands of years, providing food, materials, and spiritual significance.

All About Walruses

Scientific Name: *Odobenus rosmarus*

Type: Marine Mammal

Size: Adult males can weigh between 1,800 to 3,700 pounds (800 to 1,700 kg), while females weigh 1,200 to 2,500 pounds (600 to 1,100 kg). Walruses can reach lengths of up to 12 feet (3.6 meters) for males and 10 feet (3 meters) for females.

Diet: Primarily molluskivores, walruses feed almost exclusively on benthic bivalve mollusks. They also occasionally eat shrimp, crabs, tube worms, soft corals, and sea cucumbers.

Breeding and Young: Breeding occurs from January to March. Females give birth to a single calf after a 15 to 16-month gestation period. Calves are nursed for over a year and stay with their mothers for about two years.

Social Life and Communication: Walruses are highly social animals, forming large herds on ice or land. They communicate through a range of vocalizations and physical gestures, including bellows, grunts, and knocks.

Territory: Nomadic, following the edge of the sea ice to seasonal feeding grounds.

Sleep: 8-10 hours per day, additional long resting periods

Habitat: Icy waters of the Arctic Ocean and the subarctic seas of the Northern Hemisphere. They are found on sea ice, beaches, and shallow waters near coastlines

Lifespan: Up to 40 years in the wild

Status: Vulnerable, population is in decline

Territorial behavior is not typical of walruses; they are more focused on finding food and resting spots. They spend a lot of time on sea ice, using it as a platform to rest, molt, and give birth. When it's time to sleep, walruses can snooze in the water, staying buoyant thanks to their blubber, or on ice floes, often piling on top of each other for warmth and protection.

Walruses primarily inhabit the shallow waters of the Arctic and sub-Arctic regions, including the Chukchi Sea, Bering Sea, Laptev Sea, and the waters around Greenland and northern Canada. They prefer areas with ice cover but can also be found in open water during the summer months when the ice melts.

One fascinating fact about walruses is their ability to slow their heart rate to withstand the cold water temperatures. This adaptation, known as **bradycardia**, allows them to dive deep and stay underwater for extended periods, sometimes up to 30 minutes, while searching for food.

DID YOU KNOW?

Walrus whiskers are incredibly sensitive and can detect the smallest movements of prey on the ocean floor. They have about 400 to 700 whiskers, each filled with nerve endings that help them locate food in the murky ocean depths!

Walruses, with their distinctive appearance and remarkable adaptations, are an integral part of the Arctic marine ecosystem. These gentle giants remind us of the incredible diversity and resilience of life in the planet's coldest environments. Their social nature and unique characteristics make them one of the most interesting animals to learn about and observe in the wild.

Caribou: The Arctic's Graceful Wanderer

The caribou, also known as reindeer, roams in herds across the vast, snowy expanses of the Arctic tundra and boreal forests.

Caribou are medium to large-sized members of the deer family. Both males and females grow antlers, which is unique among deer species. Males shed their antlers after the mating season in autumn, while females keep theirs throughout the winter, using them to dig through the snow to find food. Their antlers can span up to four feet wide!

One of the most striking features of caribou is their thick coat, which consists of two layers: a dense underfur and a longer, hollow outer fur. This double layer not only keeps them warm in freezing temperatures but also helps them stay buoyant when crossing rivers and lakes. Yes, caribou are excellent swimmers, and they often have to cross bodies of water during their long **migrations**.

Caribou are **herbivores**, primarily feeding on grasses, leaves, and shrubs in the summer. In winter, when the ground is covered in snow, they rely on a special diet of lichens, also known as "reindeer moss," which they dig up with their hooves. Their hooves are uniquely adapted for this purpose, with sharp edges for cutting through ice and a spongy center that provides traction on slippery surfaces.

Breeding season, known as rutting, takes place in the fall. Males compete for females by clashing their antlers in dramatic battles. After mating, the females endure a long winter before giving birth in late spring. A single calf is usually born, and within hours, the calf is able to stand and walk. This quick adaptation is crucial for keeping up with the constantly moving herd.

Caribou are known for their migrations, which are among the longest of any land mammal. Some herds travel up to 2,700 miles (4345 km) a year, moving between

Fast Facts!

- **UV Vision:** Caribou have the unique ability to see ultraviolet light, which helps them spot things that are otherwise invisible to the human eye, like the urine trails of predators and lichen on the snow-covered ground.
- **Super Marathoners:** Caribou are known for their incredible endurance, capable of migrating over 3,000 miles in a year. That's like traveling from New York City to Los Angeles!
- **Floating Fur:** Caribou fur is made up of hollow hairs that trap air, providing excellent insulation and buoyancy. This helps them stay warm and makes them great swimmers.
- **Heat Exchangers:** Caribou have a special adaptation in their noses that warms the cold air before it reaches their lungs. Their nasal passages contain a complex network of blood vessels that exchange heat.
- **Snow Plows**: Caribou use their large, shovel-like hooves to dig through the snow to find food. Their hooves spread out wide, acting like natural snowshoes that prevent them from sinking.
- **Antler Velcro:** The velvet that covers growing antlers is rich in blood vessels and nerves, which help the antlers grow. This velvet is eventually shed, revealing the hard bone beneath.
- **Furry Walkers:** The bottom of a caribou's hooves is covered in fur, which helps protect them from the cold. This adaptation also provides good traction when walking on ice and snow.
- **Lichen Eaters:** Caribou are one of the only animals to be able to digest lichen which is one of their only available food sources in the winter. Their stomachs contain special microorganisms that allow them to extract as much energy as possible from this sparse food source.
- **Incredible Swimmers:** Caribou can swim across wide rivers and lakes, propelled by their strong legs and buoyant bodies. They often have to swim during their long migrations to reach new feeding grounds.
- **Cold Legs:** Caribou have a unique ability to regulate their body temperature. Their legs cool down more than their bodies, which helps prevent heat loss through their limbs in the frigid Arctic temperatures!

All About Caribou

Scientific Name: *Rangifer tarandus*

Type: Land Mammal

Size: Adult males weigh between 350 to 500 pounds (160 to 227 kg), and females weigh about 180 to 260 pounds (80 to 120 kg). They stand about 4 to 5 feet (1.2 to 1.5 meters) tall at the shoulder and range in length from 5.6 to 7.9 feet (1.7 to 2.4 meters).

Diet: Herbivores, feeding on a variety of vegetation including grasses, sedges, leaves, and especially lichens in the winter, known as "reindeer moss."

Breeding and Young:
Mate in late September to early November. Females give birth to a single calf after a gestation period of about 230 days, in May or June. Calves can walk shortly after birth and are weaned by autumn.

Social Life and Communication: Reindeer are highly social and form large herds for migration, which can number in the thousands. They communicate using various sounds, including grunts, bleats, and snorts, especially between mothers and calves.

Territory: They undertake extensive migrations between their winter and summer feeding grounds, which is among the longest of any land mammal.

Sleep: 6-7 hours per day spread across several naps

Habitat: They are found in the Arctic tundra and northern forests of Canada, Alaska, Greenland, Norway, Sweden, Finland, and Russia

Lifespan: 8-10 years in the wild

Status: Varies by region; some populations are threatened or endangered due to habitat loss, climate change, and overhunting their summer and winter ranges. During migration, they travel in large groups, which can number in the thousands, creating a spectacular sight as they move across the landscape.

> **DID YOU KNOW?**
> Caribou have specialized tendons in their feet that make a clicking sound when they walk. This helps herd members stay together in poor visibility conditions like heavy snow or fog!

Social animals by nature, caribou are typically quiet but do communicate using grunts, snorts, and bellows. They also rely on their keen sense of smell to find food and detect predators.

Their large, furry noses warm the cold air before it enters their lungs, and their eyes change color with the seasons—from golden in the summer to blue in the winter—improving their vision in low-light conditions.

Indigenous peoples have domesticated caribou for their meat, milk, and fur, as well as for transportation. And, of course, reindeer are famously known as Santa Claus's trusty sleigh-pullers!

Caribou, with their incredible **adaptations** and remarkable migrations, are true survivors of the Arctic wilderness. Whether roaming the tundra or featuring in holiday tales, caribou capture our imagination and remind us of the resilience and beauty of the natural world.

Ringed Seal: The Arctic's Ice-Adapted Acrobat

Beneath the vast expanse of Arctic ice, the ringed seal navigates a world of icy waters and constant challenges. These small but mighty marine mammals have perfected the art of survival in one of Earth's most unforgiving environments.

Ringed seals are compact and streamlined, typically weighing between 110 and 150 pounds (50-68 kg) and measuring about 5 feet (1.5 meters) in length. Their name comes from the distinctive pattern of dark spots surrounded by lighter rings that adorn their sleek bodies, providing excellent camouflage against the mottled ice and snow of their habitat.

These agile swimmers are remarkable divers, plunging to depths of up to 300 feet (91 meters) and staying **submerged** for as long as 45 minutes. Their diet consists primarily of small fish, shrimp, and crustaceans, which they hunt with incredible precision. Ringed seals use their sensitive whiskers, called vibrissae, to detect the slightest movements of prey in the dark, icy waters.

One of the most impressive adaptations of the ringed seal is its ability to create and maintain breathing holes in the sea ice. Using the sharp claws on their fore-flippers, they scratch away at the ice to keep these holes open throughout the winter. These breathing holes are vital for their survival, allowing them to surface for air in a landscape where the sea is often completely covered by ice.

Breeding season occurs in late winter and early spring. Female seals give birth to a single pup in a snow den built on the ice, providing shelter from extreme cold and protection from predators like polar bears and Arctic foxes. Newborn pups are born with a fluffy white coat called **lanugo**, which helps insulate them against the harsh Arctic temperatures. Within about a month, pups shed this coat for the sleek, spotted fur of an adult.

Ringed seals are generally solitary, preferring to live alone except during breeding season or when mothers are raising their pups. Despite their solitary nature, they communicate through various vocalizations, including barks, growls, and chirps, especially during mating season.

Fast Facts!

- **Deep Divers:** They can dive up to 300 feet (91 meters) deep and stay underwater for 45 minutes.
- **Whisker Hunters:** Their sensitive whiskers (vibrissae) help them detect prey movements in dark waters.
- **Camouflage Experts:** Their spotted coat provides excellent camouflage against patchy ice and snow.
- **Solitary Lifestyle:** Ringed seals are generally solitary, coming together mainly for breeding.
- **Blubber Insulation:** A thick layer of blubber keeps them warm and serves as an energy reserve.
- **Keystone Species:** They are crucial prey for polar bears, playing a vital role in the Arctic food web.
- **Climate Vulnerable:** Climate change and sea ice reduction pose significant threats to their survival.
- **Cultural Importance:** Indigenous Arctic peoples have relied on ringed seals for food and materials for millennia.
- **Ice Architects:** Ringed seals create and maintain breathing holes in sea ice using their sharp claws.
- **Excellent Swimmers:** Newborn pups can swim within hours of birth.
- **Varied Diet**: While they primarily eat fish, they also consume a variety of crustaceans and even some mollusks.
- **Widespread:** Ringed seals are the most numerous and widespread of Arctic seals.

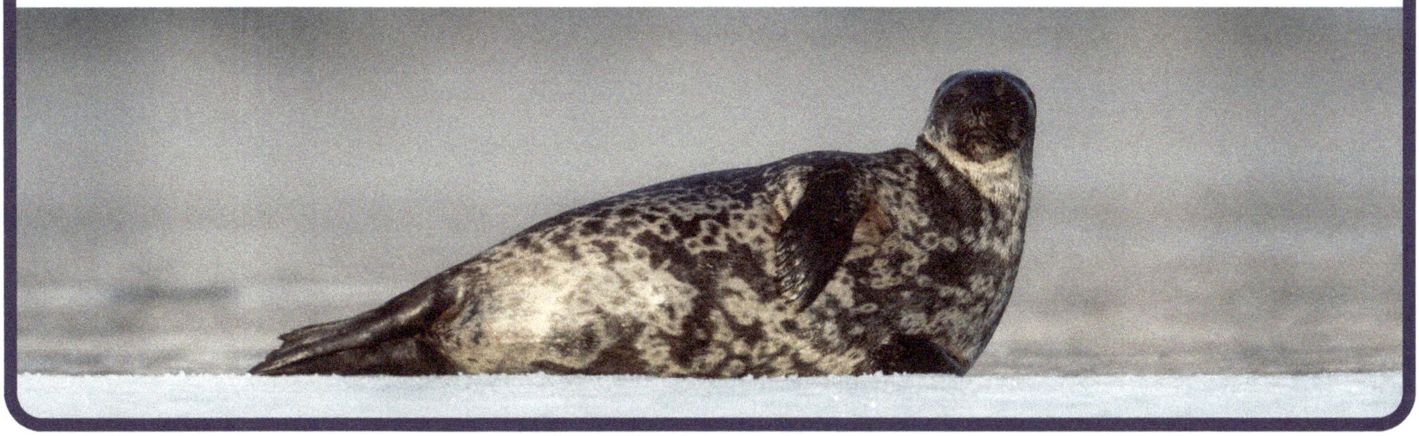

All About Ringed Seals

Scientific Name: *Pusa hispida*

Type: Marine Mammal

Size: Ringed seals are relatively small, averaging about 100-150 pounds (45-68 kg) and measuring 4-5 feet (1.2-1.5 meters) in length.

Diet: Primarily piscivorous, consuming Arctic cod, herring, and other small fish. They also eat crustaceans like shrimp and amphipods, and occasionally mollusks.

Breeding and Young: Ringed seals give birth to a single pup in snow dens on the ice, offering protection against predators and the cold. The gestation period is about 11 months, including a 3-4 month delay in implantation. Pups are born between March and April.

Social Life and Communication: Generally solitary, coming together mainly for breeding. They communicate through various vocalizations including barks, growls, and chirps, especially during mating season.

Territory: Ringed seals are circumpolar, inhabiting the Arctic Ocean and adjacent seas, including waters around Alaska, Canada, Greenland, Norway, and Russia. They're closely associated with ice-covered waters.

Sleep: Ringed seals can sleep in water or on ice. They often practice "cathemeral" sleep patterns, resting in short periods throughout the day and night.

Habitat: Ringed seals live in and around the ice-covered waters of the Arctic Ocean.

Lifespan: Up to 40 years in the wild, though 25-30 is more common.

Status: Listed as Least Concern by the IUCN, but considered vulnerable in some regions due to climate change impacts on sea ice. Some subpopulations are listed as threatened under the U.S. Endangered Species Act.

The territory of a ringed seal can vary greatly, as they follow the shifting ice and food availability. They are highly dependent on sea ice for their survival, using it for resting, breeding, and as a platform to haul out of the water. When it's time to rest, they can be seen basking on the ice, soaking up the sun's rays to stay warm.

DID YOU KNOW?

Ringed seals blood contains a high concentration of **hemoglobin**, which allows them to store more oxygen. This enables them to stay underwater for longer periods and dive deeper in search of food.

Ringed seals inhabit the Arctic and sub-Arctic regions, including the waters around Alaska, Canada, Greenland, Norway, and Russia. They are incredibly resilient, thriving in some of the harshest conditions on Earth. Their thick blubber not only provides insulation against the cold but also serves as an energy reserve during times when food is scarce.

Ringed seals play a crucial role in the Arctic ecosystem as a **keystone species**. They are the primary prey for polar bears, and the survival of these iconic Arctic predators is closely linked to ringed seal abundance. Additionally, Indigenous peoples of the Arctic have relied on ringed seals for food, clothing, and tools for thousands of years, making them culturally significant as well.

Despite their resilience, ringed seals face significant threats from climate change. The rapid reduction of sea ice due to global warming is endangering their habitat and survival. Conservation efforts are crucial to protect these remarkable creatures and preserve the delicate balance of the Arctic ecosystem.

Arctic Wolf: Resilient Hunter of the Frozen North

In the icy tundra of the Arctic, a remarkable predator roams. The Arctic wolf, a subspecies of the gray wolf, has mastered the art of survival. With its thick, white coat and piercing yellow eyes, this creature embodies the wild beauty and resilience of the far north.

Arctic wolves are slightly smaller than their gray wolf cousins. Males typically weigh between 70 and 175 pounds (32 to 79 kg), while females are somewhat lighter. Standing about 2 to 3 feet (60 to 90 cm) tall at the shoulder, they can measure up to 5 feet (1.5 meters) in length, including their bushy tails. Their compact build is an adaptation to the cold, with small, rounded ears and short snouts that minimize heat loss.

The Arctic wolf's pristine white coat is its most striking feature. This dense fur provides crucial insulation against temperatures that can plummet to -30°F (-34°C) or lower. Their large, padded paws act like natural snowshoes, allowing them to move silently across the snow without sinking.

These wolves are opportunistic hunters, their diet varying with prey availability. They primarily target caribou and musk oxen, but will also hunt Arctic hares, lemmings, and birds when larger prey is scarce. Arctic wolves are skilled hunters, often working in packs to bring down animals much larger than themselves. Their powerful jaws and sharp teeth make quick work of their prey, while their keen senses of smell, hearing, and sight help them locate food even in the challenging conditions of the polar night.

Breeding season for Arctic wolves occurs in late winter, and after a gestation period of about 63 days, the female gives birth to a litter of 2 to 6 pups in a den. These dens are often dug into the side of a hill or snowbank, providing shelter from the cold and protection from predators. Born blind and helpless, the pups rely entirely on their mother and pack for survival in their early weeks.

Fast Facts!

- **White Camouflage:** Their white fur provides perfect camouflage in snowy landscapes.
- **Cold-Adapted:** Small ears and short muzzles help minimize heat loss in freezing temperatures.
- **Pack Hunters:** They live and hunt in family groups, typically led by an alpha pair.
- **Long-Distance Travelers:** Arctic wolves can roam over 1,000 square miles (2,600 km^2) in search of food.
- **Diverse Diet:** They primarily hunt caribou and musk oxen, but also eat smaller prey like Arctic hares and lemmings.
- **Strong Swimmers:** Despite the cold waters, Arctic wolves are capable swimmers when needed.
- **Keen Senses:** Excellent hearing, sight, and smell help them locate prey in challenging Arctic conditions.
- **Long Lifespan:** They can live up to 10 years in the wild, and even longer in captivity.
- **Efficient Hunters:** Arctic wolves can survive for long periods without food, then consume up to 20 pounds (9 kg) of meat in a single meal.
- **Vocal Communicators:** They use a variety of howls, barks, and growls to communicate within their pack.
- **Thick Coat:** Their double-layered fur can be up to 2 inches (5 cm) thick, providing crucial insulation.
- **No Fear of Humans:** Due to limited contact, Arctic wolves show little fear of humans compared to other wolf subspecies.
- **Den Diggers:** They create dens in snow or soil for shelter and raising pups.
- **Efficient Metabolism:** Arctic wolves have adapted to extract maximum nutrition from their prey, crucial for survival in their harsh environment.

All About Arctic Wolves

Scientific Name: *Canis lupus arctos*

Type: Mammal

Size: Adult Arctic wolves typically weigh between 70 to 125 pounds (32 to 57 kg) and can measure up to 5 feet (1.5 meters) in length, including their tail.

Diet: Carnivorous, primarily preying on muskoxen, Arctic hares, and caribou. They also eat lemmings, other rodents, and birds when larger prey is scarce.

Breeding and Young: Mating season occurs once a year in late winter or early spring. After a gestation period of about 63 days, females give birth to 2-6 pups on average in dens dug into snow-banks or soil.

Social Life and Communication: Live and hunt in packs typically consisting of 5 to 10 members, led by an alpha pair. They communicate through howls, growls, and body language to maintain pack structure and territory.

Territory: Inhabit vast, circumpolar Arctic regions, roaming large territories that can span 600 to 1,000 square miles (1,000 to 2,600 square kilometers) to find food and shelter.

Sleep: Arctic wolves have adaptive sleeping habits, able to rest in a variety of conditions. They often sleep in the open, curling up with their tails over their noses for warmth.

Habitat: Primarily found in the Arctic Circle, encompassing the northern parts of North America and Greenland, where they are adapted to living in cold environments with permanent ice and snow.

Lifespan: Can live up to 7-10 years in the wild, though longer in captivity.

Status: Classified as Least Concern by the IUCN. Climate change poses significant challenges to their habitat and food sources, potentially impacting their long-term survival.

Arctic wolves are highly social animals, living and hunting in packs that typically range from 6 to 20 members. Each pack is led by an **alpha** male and female, with a strict hierarchy governing pack life. Communication is key to their survival, and Arctic wolves use a variety of vocalizations, body language, and scent marking to convey information within the pack and to other wolves.

DID YOU KNOW?

Arctic wolves have a special adaptation in their paws that allows them to walk on ice without slipping. The pads of their feet have small bumps called papillae that increase friction and provide better grip on slippery surfaces.

The territory of an Arctic wolf pack can span hundreds of square miles. These nomadic predators follow the migrations of their prey, rarely staying in one place for long. When resting, pack members often curl up together, using their thick tails to cover their faces and conserve heat.

Found throughout the Arctic regions of North America and Greenland, Arctic wolves inhabit some of the most remote and inhospitable places on Earth. Their ability to thrive in these extreme conditions is a testament to their remarkable adaptations and resilience.

Arctic wolves play a crucial role in their **ecosystem** by controlling prey populations, which in turn helps maintain the delicate balance of Arctic **biodiversity**. However, like many Arctic species, they face an uncertain future due to climate change. As temperatures rise and sea ice melts, the habitats and migration patterns of their prey are changing, presenting new challenges for these adaptable predators.

Snowy Owl: The Arctic's Silent Sentinel

Imagine a large, ghostly bird with piercing yellow eyes, scanning the frozen ground for prey. The snowy owl, one of the most striking and resilient birds of the Arctic, is perfectly adapted to life in this harsh environment.

Snowy owls are among the largest owl species in North America. Their impressive size is matched by an equally impressive wingspan, stretching up to 4.8 feet (146 cm). From head to tail, they measure about 20-28 inches (51-71 cm). Males are typically smaller and whiter than females, who display more dark barring on their feathers, providing better camouflage during nesting.

These owls are built for the cold. Their entire body, including their legs and toes, is covered in dense, insulating feathers. This **plumage** not only keeps them warm in temperatures as low as -40°F (-40°C) but also muffles the sound of their flight, making them nearly silent hunters.

The diet of snowy owls is primarily composed of lemmings, small rodents that can make up 90% of their diet in some areas. During peak lemming years, a single owl might consume 1,600 lemmings or more. When lemmings are scarce, these versatile predators will hunt a variety of prey, including Arctic hares, ptarmigan, ducks, and even fish.

Snowy owls are diurnal, meaning they're active during the day, unlike most owl species. This adaptation allows them to hunt during the long daylight hours of the Arctic summer. Their large, yellow eyes are incredibly sensitive, capable of spotting movement from over half a mile away. Combined with their acute hearing, which can detect prey under thick snow, snowy owls are formidable hunters in any conditions.

Fast Facts!

- **Arctic Specialists:** Adapted to survive in temperatures as low as -40°F (-40°C).
- **Impressive Size:** One of the heaviest owl species in North America, weighing up to 6.6 pounds (3 kg).
- **Daylight Hunters:** Unlike most owls, snowy owls are active during the day.
- **Lemming Lovers:** Can eat up to 1,600 lemmings in a single year when prey is abundant.
- **Ground Nesters:** Build their nests on the tundra floor, often on slightly elevated areas.
- **Long-distance Travelers:** Some individuals migrate thousands of miles during winter "irruptions."
- **Silent Flight:** Special wing feathers allow them to fly almost noiselessly.
- **Excellent Vision:** Can spot movement from over half a mile away.
- **Feathered Feet:** Even their toes are covered in insulating feathers.
- **Varied Diet:** While they prefer lemmings, they'll eat a wide range of prey, including fish.
- **Long-lived:** Can survive up to 10 years in the wild, and over 20 in captivity.
- **Sexual Dimorphism:** Females are typically larger and darker than males.
- **Circumpolar Range:** Found across the northern regions of North America, Europe, and Asia.
- **Climate Vulnerable:** Their Arctic habitat is increasingly threatened by global warming.
- **Cultural Icon:** Featured in many Indigenous cultures' stories and beliefs.
- **Rotating Head:** Snowy owls can rotate their heads up to 270 degrees, allowing them to see in almost any direction without moving their bodies.
- **Asymmetrical Ears:** Their ears are positioned at different heights on their head, helping them pinpoint the exact location of sounds in 3D space.

All About Snowy Owls

Scientific Name: *Bubo scandiacus*

Type: Birds (Aves)

Size: Adult snowy owls have a wingspan of about 4.2 to 4.8 feet (1.3 to 1.5 meters) and weigh between 3.5 to 6.6 pounds (1.6 to 3 kg), with females generally larger than males.

Diet: Carnivorous, primarily preying on lemmings and voles. Their diet also includes other small mammals, birds, and occasionally fish.

Breeding and Young: Breeding season typically occurs from May to June. Females lay 3-11 eggs, depending on food abundance, with an incubation period of about 32 days.

Social Life and Communication: Generally solitary outside of the breeding season. They use a variety of vocalizations including hoots and hisses, as well as bill-clapping for communication, especially during mating season and territorial defense.

Territory: Snowy owls are nomadic, with territory sizes varying greatly depending on the availability of prey. During breeding season, they can defend territories of up to 3.9 square miles (10 square kilometers).

Sleep: Diurnal (active during the day), unlike most owl species. They are adapted to the Arctic's continuous daylight in summer and can be active at any time.

Habitat: Inhabit open Arctic tundra during breeding season. Winter habitats can extend into southern Canada and northern United States, as well as parts of Eurasia, including open fields, marshes, and coastal areas.

Lifespan: Can live up to 9-10 years in the wild, with recorded lifespans of over 20 years in captivity.

Status: Listed as Vulnerable by the IUCN due to population declines, largely attributed to climate change impacts on their Arctic breeding grounds.

Breeding season for snowy owls begins in May. Unlike tree-nesting owls, they build their nests directly on the ground, often on a slight rise for a better vantage point. Females lay 3-11 eggs, with clutch size depending on food availability. **Incubation** lasts about 32 days, with the female staying on the nest while the male hunts and defends the territory. The chicks, covered in fluffy white down at hatching, grow quickly and can leave the nest on foot at about 25 days old, though they don't fledge until 7 weeks.

DID YOU KNOW?

Snowy owls have a third eyelid that acts like sunglasses, protecting their eyes from bright sunlight reflecting off snow while still allowing them to see clearly.

While generally solitary, snowy owls may gather in areas of abundant prey during winter. Their migrations are irregular and driven by food availability, a behavior known as **"irruption."** Some individuals remain in the Arctic year-round, while others may travel thousands of miles south in search of food.

Snowy owls are found across the circumpolar north, inhabiting the Arctic regions of North America, Europe, and Asia. During winter, they may venture as far south as the northern United States, sometimes even appearing in unlikely urban settings.

These magnificent birds face several challenges, including climate change, which affects their Arctic habitat and prey populations. Collisions with human structures during their southern migrations also pose significant risks.

Narwhal: The Arctic's Enigmatic Unicorn

In the frigid waters of the Arctic, a creature of legend glides through the icy depths. The narwhal, often called "the unicorn of the sea," captivates our imagination with its long, spiral tusk and mysterious nature.

Narwhals are medium-sized whales, members of the family that includes belugas. Males typically measure about 13-15 feet (4-4.5 meters) long, with females slightly smaller at 11-13 feet (3.5-4 meters). They can weigh between 1,760-3,530 pounds (800-1,600 kg), with males generally larger than females.

The narwhal's most distinctive feature is its tusk, which is actually an elongated canine tooth that spirals counterclockwise up to 8.8 feet (2.7 meters) long. Primarily found in males, this tusk has about 10 million nerve endings and is thought to play a role in sensing environmental conditions and possibly in mating rituals.

Narwhals are supremely adapted to Arctic life. Their diet consists mainly of Greenland halibut, Arctic and polar cod, squid, and shrimp. They are incredible divers, plunging to depths of up to 5,900 feet (1,800 meters) and staying submerged for up to 25 minutes. Their echolocation abilities help them navigate and hunt in the dark Arctic waters.

These whales have a thick layer of blubber, up to 4 inches (10 cm) thick, which insulates them from the frigid Arctic waters and serves as an energy reserve. They can also slow their heart rate during deep dives, conserving oxygen and allowing for extended foraging trips.

Breeding occurs in spring, with females giving birth to a single calf every 2-3 years after a 14-month gestation period. Calves are born in summer in coastal waters, where they're protected and nurtured. They're weaned after about

Fast Facts!

- **Tusk Sensors:** Their tusks have millions of nerve endings, possibly used to sense water temperature, pressure, and salinity.
- **Deep Divers:** Can dive almost 1.2 miles (1.9 km) deep, deeper than most marine mammals.
- **Blubber Blanket:** Their blubber layer makes up 40% of their body weight.
- **Toothless Hunters:** Aside from their tusks, narwhals have no other teeth and swallow their prey whole.
- **Pod Life:** Live in pods of 5-10, but can gather in groups of thousands during migration.
- **Long-lived:** Can live up to 50 years in the wild.
- **Slow Growers:** Females reach sexual maturity at about 8 years, males at about 12 years.
- **Ice Navigators:** Use leads (channels) in sea ice to breathe and travel.
- **Echolocation Experts:** Produce high-pitched clicks for navigation and finding prey.
- **Cultural Icon:** Featured in Inuit art and legend for centuries.
- **Limited Range:** Found only in Arctic waters, primarily around Greenland and northern Canada.
- **Climate Vulnerable:** Their specialized Arctic habitat is rapidly changing due to global warming.
- **Tusked Females:** While rare, some females do develop tusks.
- **Seasonal Feast:** Consume large amounts of prey in summer to build up blubber for winter.
- **Silent Swimmers:** Despite their size, narwhals are nearly silent swimmers, an adaptation for avoiding predators like orcas.

All About Narwhals

Scientific Name: *Monodon monoceros*

Type: Marine Mammal

Size: Adult narwhals typically grow to 13 to 18 feet (4 to 5.5 meters) in length, including the tusk, and weigh between 1,760 to 3,530 pounds (800 to 1,600 kg), with males generally larger than females.

Diet: Carnivorous, primarily feeding on Greenland halibut, Arctic and polar cod, squid, and shrimp found in Arctic waters.

Breeding and Young: Breeding occurs in spring. Females give birth to a single calf every 2 to 3 years after a gestation period of about 14 months.

Social Life and Communication: Narwhals are social animals, living in groups called pods that typically consist of 5 to 10 individuals, but can sometimes form larger groups of hundreds or even thousands during migration. They communicate using a variety of clicks, whistles, and pulsed sounds.

Territory: Nomadic and migratory, narwhals travel in Arctic waters according to seasonal changes, moving closer to coastlines in summer and into deeper, ice-covered offshore waters in winter.

Sleep: Narwhals, like other cetaceans, are believed to sleep unihemispherically, meaning one half of their brain remains alert while the other half sleeps, allowing them to surface for air and remain aware of their surroundings.

Habitat: Primarily found in the Arctic waters around Greenland, Canada, and Russia, often associated with pack ice.

Lifespan: Can live up to 50 years in the wild, with some estimates suggesting they may live even longer.

Status: Listed as Least Concern by the IUCN as of 2017, but facing increasing threats from climate change affecting sea ice habitat, increased human activity in the Arctic, and potential predation by orcas.

20 months but may stay with their mothers for several years.

Narwhals are highly social, typically found in pods of 5-10 individuals, though gatherings of hundreds or even thousands have been observed during migrations. They communicate using a complex array of clicks, whistles, and pulsed sounds, which help in navigation, finding food, and social bonding.

DID YOU KNOW?

Narwhals can change color with age. Young narwhals are blue-gray, adults are mottled gray, and older narwhals can be almost completely white.

Found primarily in the Arctic waters around Greenland, Canada, and Russia, narwhals inhabit both deep offshore areas and shallow coastal waters. They migrate seasonally, moving between summer feeding grounds in coastal areas and wintering grounds in deeper, ice-covered offshore waters.

Despite their iconic status, narwhals face significant threats. Climate change is altering their Arctic habitat, affecting ice patterns and prey availability. Increased human activity in the Arctic, including shipping and oil exploration, poses risks of noise pollution and potential oil spills.

Arctic Hare: The Tundra's Resilient Hopper

Arctic hares are small, white-furred mammals that live in the cold, snowy regions of the Arctic. They are well adapted to survive the extreme cold with their thick fur and powerful hind legs, which help them hop quickly over the snow.

Arctic hares are larger than their southern cousins, measuring about 22 to 28 inches (55 to 70 cm) in length and weighing between 6.6 to 11 pounds (3 to 5 kg). Their most striking feature is their seasonal camouflage. In winter, their fur is pure white, blending seamlessly with the snow-covered landscape. As summer approaches, their coat transforms to a blue-gray or brown, matching the rocky tundra.

These hares are built for speed and endurance in the Arctic environment. Their large, fur-covered feet act as natural snowshoes, distributing their weight to prevent sinking into soft snow. When danger threatens, Arctic hares can reach speeds up to 40 mph (64 km/h) in short bursts, using their powerful hind legs to bound across the tundra.

The diet of Arctic hares varies with the seasons. In summer, they feast on a variety of plants, including leaves, buds, berries, and flowers. Winter presents a greater challenge, forcing them to dig through snow to find woody plants, mosses, and lichens. Their strong incisors are perfectly adapted for gnawing on tough Arctic vegetation.

To survive the extreme cold, Arctic hares have several adaptations. Their compact body shape, short ears, and small nose minimize heat loss. A thick layer of fur, which is hollow and provides excellent insulation, covers their entire body, including their feet. In particularly harsh weather, they may huddle together in groups, forming a tight circle to conserve warmth.

Breeding season for Arctic hares begins in late April or early May. After a gestation

Fast Facts!

- **Champion Jumpers:** Can leap up to 7 feet (2.1 m) in a single bound.
- **Nocturnal Nature:** Most active during twilight and night hours.
- **Built-in Snowshoes:** Large, fur-covered feet prevent sinking in snow.
- **Efficient Eaters:** Can survive on woody plants and lichens in winter.
- **Quick Growers:** Leverets reach adult size in just 5 months.
- **Long-lived:** Can survive up to 5 years in the wild.
- **Social Gatherings:** Form large groups called "downs" for protection.
- **Keen Senses:** Large eyes positioned high on the head provide excellent vision.
- **Cold-resistant:** Can withstand temperatures as low as -40°F (-40°C).
- **Camouflage Experts:** Change fur color twice a year to match their environment.
- **Silent Communicators:** Use body language more than vocalizations.
- **Constant Grazers:** Spend up to 20 hours a day feeding in summer.
- **Predator Prey:** Important food source for Arctic foxes, wolves, and birds of prey.
- **No Hibernation:** Remain active year-round despite harsh winter conditions.
- **Climate Sensitive:** Vulnerable to the effects of global warming on Arctic ecosystems.
- **Thermoregulation Experts:** Arctic hares can restrict blood flow to their extremities, concentrating warmth in their core body to survive extreme cold.
- **UV Protection:** Their eyes have a reflective layer that helps protect them from the intense UV radiation reflected by snow and ice in their environment.

All About Arctic Hares

Scientific Name: *Lepus arcticus*

Type: Mammal

Size: Adult Arctic hares typically measure 22 to 28 inches (55 to 70 cm) in length and weigh between 6.6 to 11 pounds (3 to 5 kg), with females generally slightly larger than males.

Diet: Herbivorous, primarily feeding on woody plants, mosses, and lichens in winter; switches to buds, berries, leaves, flowers, and grasses in summer.

Breeding and Young: Breeding season is in late April to early May. Females give birth to one to eight leverets (typically 2-5) after a gestation period of about 50 days.

Social Life and Communication: Can be solitary but often form large groups, especially in winter. These groups, sometimes called "downs," can number up to 200 individuals for better protection against predators and to conserve heat.

Territory: They are highly adaptable, living in the tundra and icy environments of the Arctic regions, with some seasonal migrations for food.

Sleep: Arctic hares rest in shallow depressions called "forms" and can remain vigilant even while resting, though they don't necessarily sleep with their eyes open.

Habitat: Inhabits the Arctic tundra and rocky areas of northern Canada, Greenland, and some Arctic islands.

Lifespan: Can live up to 5 years in the wild, though predation and harsh conditions often shorten this lifespan.

Status: Listed as Least Concern by the IUCN, but climate change and habitat disruption pose potential long-term risks to their populations.

period of about 50 days, females give birth to a litter of two to eight leverets. Unlike rabbits, Arctic hare leverets are born fully furred with open eyes, capable of hopping within hours of birth. This **precocial** development is crucial for survival in the unforgiving Arctic environment.

While often solitary, Arctic hares can form large groups, especially in winter. These gatherings, which can number up to 200 individuals, provide protection from predators and help conserve heat. Communication within these groups involves a variety of vocalizations and body language.

DID YOU KNOW?

Arctic hares don't dig burrows like rabbits do. Instead, they create shallow depressions in the snow or ground called "forms" where they rest and seek shelter from the wind.

Arctic hares inhabit the tundra regions of Greenland and the Canadian Arctic islands. Their range extends from Newfoundland and Labrador in the east to Alaska in the west. They are well-adapted to life on the treeless tundra, where they use rock crevices and shallow depressions for shelter.

Despite their adaptations, Arctic hares face challenges from climate change. Warming temperatures are altering plant communities and predator-prey dynamics in the Arctic, potentially impacting the hares' food sources and exposing them to new predators.

Puffin: The Charming Clown of the Arctic Seas

The puffin is a striking bird with its colorful beak and black and white plumage. The puffin, often referred to as the "clown of the sea," is perfectly adapted to its maritime environment.

Puffins are small seabirds, standing about 10 inches (25 cm) tall with a wingspan of 20 to 24 inches (50-60 cm). Their most distinctive feature is their large, triangular beak, which becomes brightly colored during the breeding season with vibrant shades of orange, yellow, and blue. This colorful outer layer is shed after breeding, leaving a smaller, duller beak for the winter months.

These birds are extraordinary swimmers, using their wings to "fly" underwater in pursuit of fish. Puffins can dive to depths of up to 200 feet (60 meters) and stay submerged for up to a minute. Their specialized beaks are designed to hold multiple fish at once, lined up crosswise, allowing them to carry up to 60 small fish in a single trip back to their chicks.

On land, puffins may appear somewhat clumsy, waddling on short legs. However, they are strong fliers, capable of beating their wings up to 400 times per minute and reaching speeds of 55 mph (88 km/h). This flying ability is crucial for their long migrations between breeding grounds and wintering areas at sea.

Breeding season brings puffins back to coastal cliffs and islands where they nest in large colonies. They are monogamous, often returning to the same burrow with the same mate year after year. Puffins dig these burrows using their sharp claws and beaks, creating a safe space for their single egg. Both parents share incubation duties and care for the chick, or "puffling," which stays in the burrow for about 6 weeks before heading out to sea.

Fast Facts!

- **Colorful Beaks:** Their beaks are brightest during breeding season and shed outer layers in winter.
- **Deep Divers:** Can plunge up to 200 feet (60 m) underwater in search of fish.
- **Fish Carriers:** Able to hold multiple fish in their beaks at once, thanks to special tongue adaptations.
- **Monogamous Mates:** Often return to the same partner and nesting site each breeding season.
- **Burrow Nesters:** Dig nesting burrows in soil or use rock crevices on coastal cliffs.
- **Single Egg Layers:** Typically lay only one egg per year, which both parents incubate.
- **Swift Flyers:** Can beat their wings up to 400 times per minute when flying.
- **Long-lived:** Can survive for 20 years or more in the wild.
- **Winter Wanderers:** Spend most of the year alone at sea, returning to land only to breed.
- **Colony Dwellers:** Nest in large groups, sometimes numbering thousands of birds.
- **Specialized Swimmers:** Use their wings to “fly” underwater when hunting fish.
- **Climate Vulnerable:** Changing ocean temperatures affect their food supply and breeding success.
- **Night Travelers:** Pufflings leave their nests at night to avoid predators on their first journey to sea.
- **Excellent Navigators:** Can find their way back to the same nesting site after months at sea.
- **Cultural Icons:** Featured in various cultures’ folklore and modern popular media.

All About Puffins

Scientific Name: *Fratercula arctica*

Type: Bird (Seabird)

Size: Adult Atlantic puffins typically measure about 10 to 12 inches (25 to 30 cm) in length and weigh around 12 to 17 ounces (340 to 480 grams), with wingspans of 20 to 24 inches (50 to 60 cm).

Diet: Primarily piscivorous, feeding on small fish such as sand eels, capelin, and herring. They also eat crustaceans and mollusks.

Breeding and Young: Breeding season begins in late spring. Puffins nest in burrows or crevices on cliff sides, laying a single egg per year, which both parents incubate for about 39-43 days.

Social Life and Communication: Highly social during breeding season, puffins form large colonies on coastal cliffs. They communicate with a variety of vocalizations and displays.

Territory: Exhibit high site fidelity, often returning annually to the same breeding site and exact burrow. They defend their immediate nesting area but are not strongly territorial beyond this.

Sleep: Puffins sleep on the water's surface when at sea, and can also be found resting in their burrows during the breeding season.

Habitat: Nest in coastal colonies in the North Atlantic and Arctic Oceans, from northern Europe to Canada and the northeastern United States.

Lifespan: Can live up to 20-25 years in the wild, with survival rates increasing after their first year.

Status: Listed as Vulnerable by the IUCN due to rapid population declines in parts of their range. Threats include climate change affecting food supply, pollution, introduced predators on breeding islands, and bycatch in fishing gear.

Puffins are social birds, engaging in various behaviors to communicate within their colonies. They use a range of vocalizations, including growls and grunts, especially during the breeding season. This social structure helps protect them from predators and ensures better survival rates for their young.

DID YOU KNOW?
Puffins have special ridges on their beaks and tongues that allow them to hold up to 60 small fish at once while still being able to catch more!

Found across the North Atlantic, puffins inhabit the coastlines and islands of countries including Iceland, Norway, Greenland, Newfoundland, and parts of the UK. They are well-adapted to cold conditions, with waterproof feathers providing insulation against frigid temperatures.

One of the puffin's most impressive abilities is its **navigation**. After spending months at sea, they can find their way back to the exact same nesting site with remarkable accuracy, a crucial skill for their survival and the continuity of their breeding colonies.

Puffins, with their distinctive looks and remarkable adaptations, are truly captivating seabirds. Whether diving skillfully for fish or nesting on rocky cliffs, puffins enchant anyone lucky enough to observe them in their natural habitat.

Musk Ox: The Arctic's Woolly Warrior

The musk ox is a large, shaggy creature with long, curved horns and a thick coat that hangs down to its legs, providing essential insulation against the freezing temperatures. Musk oxen are perfectly adapted to their environment, thriving where few other animals can survive.

Musk oxen are large, sturdy animals that have roamed the Arctic since the Pleistocene era. Adult males, called bulls, can weigh up to 900 pounds (400 kg) and stand about 5 feet (1.5 meters) tall at the shoulder. Females, or cows, are slightly smaller, typically weighing between 500-700 pounds (225-320 kg).

Their most distinctive feature is their thick, woolly coat, which consists of long, coarse guard hairs and a dense undercoat called qiviut. This qiviut is one of the warmest natural fibers in the world, eight times warmer than sheep's wool and finer than cashmere. This exceptional insulation allows musk oxen to withstand temperatures as low as -70°F (-57°C).

Musk oxen are herbivores, grazing on grasses, sedges, and woody plants during the short Arctic summer. In winter, they use their strong hooves to dig through snow, accessing mosses and lichens beneath. Their specialized digestive system allows them to extract maximum nutrition from this sparse, fibrous diet.

These animals are highly social, living in herds that typically range from 10 to 20 individuals, though larger groups may form during calving or mating seasons. This social structure is crucial for their survival, providing protection against predators like wolves and polar bears.

When threatened, musk oxen display one of nature's most impressive defensive

Fast Facts!

- **Ice Age Survivors:** Musk oxen have roamed the Arctic tundra ifor over 200,000 years.
- **Two-Layered Coat:** They have a unique coat with a long, coarse outer layer and a soft, warm undercoat called qiviut, which is shed in spring and highly valued for its warmth and softness.
- **Fortress Formation:** Form a defensive circle with horns facing outward when threatened.
- **Extreme Cold Tolerance:** Can survive temperatures as low as -70°F (-57°C).
- **Strong Diggers:** Use their hooves to dig through snow for food in winter.
- **Dramatic Mating Battles:** During the mating season, males engage in head-to-head combat to win over females, leading to dramatic clashes that can be heard from far away.
- **Precocial Calves:** Newborns can stand and follow the herd within hours of birth.
- **Ecosystem Role:** Their grazing habits help to shape the Arctic ecosystem, promoting the growth of certain types of vegetation.
- **Thick Hooves:** Their large, splayed hooves act like built-in snowshoes, helping them walk on snow and dig for food.
- **Excellent Swimmers:** Despite their bulky appearance, musk oxen are surprisingly good swimmers and can cross rivers and lakes.
- **Ancient DNA:** Musk oxen have low genetic diversity due to near-extinction events in their past, making them a subject of interest for geneticists.
- **Herd Hierarchy:** Musk ox herds have a clear social structure, with older females often leading the group.
- **Specialized Eyes:** Their eyes have rectangular pupils, similar to goats, giving them a wide field of vision to spot predators.
- **Weather Predictors:** Indigenous peoples have traditionally used musk ox behavior to predict weather changes in the Arctic.

All About Musk Oxen

Scientific Name: *Ovibos moschatus*

Type: Mammal

Size: Adult musk oxen typically stand about 4 to 5 feet (1.2 to 1.5 meters) at the shoulder. Males can weigh between 400 to 900 pounds (180 to 410 kg), while females generally weigh 500 to 700 pounds (225 to 320 kg).

Diet: Herbivorous, feeding primarily on grasses, mosses, willows, and other tundra vegetation.

Breeding and Young: Breeding season occurs in late summer or early fall. Females give birth to a single calf every two years after a gestation period of about 8 to 9 months.

Social Life and Communication: Live in mixed-sex herds that typically range from 10 to 20 individuals, providing protection against predators and harsh conditions. They communicate using grunts, bellows, and a distinctive "roar-snort" during mating season.

Territory: Musk oxen are not strictly territorial but maintain home ranges that vary in size depending on food availability and season. They may move considerable distances to find better grazing areas.

Sleep: Musk oxen rest and sleep lying down, often taking turns within the herd to remain alert for predators. They may also rest standing up, especially in severe weather.

Habitat: Found in the Arctic tundra across northern Canada, Greenland, and Alaska, with introduced populations in Norway, Sweden, and Siberia. They prefer open, cold habitats with access to grazing areas.

Lifespan: Can live up to 20-25 years in the wild, though 12 to 15 years is more common.

Status: Listed as Least Concern by the IUCN due to stable overall population, but some local populations are vulnerable to climate change and hunting pressures.

strategies. The herd forms a tight circle, with adults facing outward and calves protected in the center. This "fortress formation" presents would-be attackers with an intimidating wall of sharp horns and powerful bodies.

DID YOU KNOW?

A musk ox's horns never stop growing! The horns grow rings each year, similar to tree rings, which scientists can use to determine the age of the animal.

Breeding occurs in late summer, with bulls engaging in fierce competitions for mating rights. These contests involve dramatic head-to-head clashes that can be heard echoing across the tundra. After a gestation period of about 8 months, cows give birth to a single calf in the spring. Calves are born with a fuzzy brown coat and can stand within hours of birth, a necessary adaptation for survival in the harsh Arctic environment.

Musk oxen communicate through a variety of vocalizations, including grunts, bellows, and a unique "roar-snort" used during mating season. They also use scent to communicate, with bulls producing a strong musky odor during breeding season, which gives the species its name.

Found in the Arctic regions of North America and Greenland, with introduced populations in Norway, Sweden, and Siberia, musk oxen play a crucial role in their ecosystem. Their grazing habits help shape Arctic vegetation, and they serve as an important food source for indigenous communities.

Despite their adaptations, musk oxen face challenges from climate change, which is altering their habitat and food availability. Conservation efforts are crucial to ensure the survival of these Ice Age relics in a rapidly changing Arctic.

Beluga Whale: The Arctic's Melodic Wanderer

In the icy waters of the Arctic and sub-Arctic regions, the beluga whale, glides gracefully beneath the surface. Beluga whales are renowned for their vocal abilities and social nature, making them one of the most fascinating marine mammals in their chilly habitat.

Beluga whales are medium-sized cetaceans, growing up to 13-20 feet (4-6.1 meters) in length and weighing between 1,100-3,300 pounds (500-1,500 kg). Their most striking feature is their pure white coloration, which provides excellent camouflage in their icy habitat and has earned them the nickname "white whales."

These whales are easily recognized by their distinctive bulbous forehead, called a melon, which they can change shape at will. This unique ability aids in their impressive vocal repertoire and gives them an almost human-like expressiveness. Unlike most whales, belugas have a flexible neck, allowing them to turn their heads in all directions—a useful adaptation for navigating through ice-filled waters.

Belugas are opportunistic feeders with a diverse diet that includes various fish species, squid, octopus, and crustaceans. They use **echolocation** to hunt, emitting a series of clicks that bounce off objects in the water, helping them locate prey in the often murky Arctic waters. Their adaptability allows them to hunt in both shallow coastal areas and depths of up to 1,000 feet (300 meters).

The vocal abilities of beluga whales are truly remarkable. They produce a wide array of sounds including whistles, squeals, moos, chirps, and trills, earning them their "canary of the sea" moniker. These vocalizations serve multiple purposes: communication

Fast Facts!

- **White Camouflage:** Their white color provides excellent Arctic camouflage.
- **Echolocation Experts:** Use sound waves to navigate and find prey in murky waters.
- **Flexible Necks:** Can turn their heads in all directions, unlike most whales.
- **Shallow Water Lovers:** Often found in coastal areas and even swim up rivers.
- **Annual Molters:** Shed their outer skin layer every summer.
- **Long-lived:** Can survive up to 50 years in the wild.
- **Melodious Communicators:** Produce a wide range of vocalizations for different purposes.
- **Thick Blubber:** Their blubber layer can be up to 5 inches (12 cm) thick for insulation.
- **Slow Swimmers:** Typically swim at 2-6 mph, but can reach 14 mph in short bursts.
- **No Dorsal Fin:** Lack a dorsal fin, an adaptation for swimming under ice.
- **Varied Diet:** Eat over 100 different types of primarily bottom-dwelling sea creatures.
- **Delayed Maturity:** Females don't reach sexual maturity until about 9 years old.
- **Pod Life:** Live in family groups called pods, sometimes forming large seasonal gatherings.
- **Pressure-Resistant:** Belugas can dive to depths of over 3,000 feet (900 meters) thanks to their pressure-resistant bodies.
- **Excellent Hearing:** They have exceptional hearing abilities, with a hearing range about seven times wider than that of humans.
- **Habitat Pioneers:** Belugas are often the first whales to return to ice-covered Arctic waters in spring.
- **Temperature Adaptations:** Their circulatory system can adjust blood flow to maintain body temperature in frigid waters.
- **Social Learners:** Young belugas learn hunting techniques and migration routes from their elders in the pod.

All About Beluga Whales

Scientific Name: *Delphinapterus leucas*

Type: Marine Mammal

Size: Adult beluga whales typically reach lengths of 13 to 20 feet (4 to 6 meters) and weigh between 1,100 to 3,300 pounds (500 to 1,500 kg), with males slightly larger than females.

Diet: Carnivorous, primarily feeding on fish, crustaceans, and cephalopods found in Arctic waters.

Breeding and Young: Breeding occurs in late winter to early spring. Females give birth to a single calf every 2 to 3 years after a gestation period of about 14 to 15 months. Calves are nursed for about two years.

Social Life and Communication: Highly social, belugas live in pods that can range from a few individuals to hundreds, known for their complex vocalizations and use of echolocation for navigation and communication.

Territory: Belugas migrate seasonally, moving between summer estuaries and open sea ice waters in winter for feeding and breeding. Some populations can travel thousands of miles annually.

Sleep: Beluga whales engage in unihemispheric slow-wave sleep, where one half of the brain remains alert while the other half sleeps. This allows them to continue swimming and surfacing to breathe while resting.

Habitat: Primarily found in Arctic and sub-Arctic waters, including coastal bays, estuaries, and deep offshore waters.

Lifespan: Can live up to 50 years in the wild, with sexual maturity reached at about 8-9 years for females and 12-14 years for males.

Status: Listed as Least Concern globally by the IUCN, but some subpopulations are considered endangered or threatened due to various factors including pollution, habitat loss, noise disturbance, and climate change.

within their pods, echolocation for hunting and navigation, and possibly even to stun prey. Highly social animals, belugas typically live in pods ranging from a few individuals to several hundred. These groups often consist of related individuals and demonstrate strong social bonds. During summer months, belugas gather in large numbers in shallow estuaries and bays, not only to feed but also to engage in an annual molt, rubbing against the gravel bottom to shed their old skin.

Breeding occurs in spring, with females giving birth to a single calf every two to three years after a 14-15 month gestation period. Calves are born dark gray and gradually lighten to white as they mature over about five years. Mothers are devoted caregivers, nursing their young for up to two years and teaching them essential survival skills.

Belugas inhabit the Arctic and sub-Arctic waters around North America, Russia, and Greenland. They are well-adapted to life among sea ice, using their rounded foreheads to break through thin ice to create breathing holes. During winter, they often remain near the ice edge, moving to coastal areas and estuaries in summer.

DID YOU KNOW?

Beluga whales can mimic a variety of sounds, including human speech patterns. In captivity, some belugas have even learned to imitate words!

Despite their adaptability, beluga whales face numerous threats including climate change, which is altering their Arctic habitat, as well as pollution and human disturbances. Some populations are considered endangered, highlighting the need for ongoing conservation efforts to protect these enchanting marine mammals.

Spotlight: Lemming

The lemming, a tiny rodent about the size of a hamster, plays a huge role in the Arctic ecosystem. These furry little animals are a crucial food source for many Arctic predators, including foxes, owls, and even polar bears. Their population booms and busts help keep the entire Arctic food web in balance, making them a true keystone species of the tundra.

About Lemmings:

- **Scientific name:** *Lemmus lemmus*
- **Size:** 3-6 inches long (7-15 cm)
- **Weight:** 1-4 ounces (30-110 g)
- **Lifespan:** 1-2 years in the wild
- **Diet:** Mainly plants, mosses, and lichens
- **Habitat:** Arctic tundra across North America and Eurasia
- **Fur:** Changes color seasonally - brown in summer, white in winter
- **Hibernation:** Don't hibernate; remain active year-round
- **Skills:** Excellent swimmers and diggers
- **Tunneling:** Can tunnel through snow to find food and shelter
- **Population:** Explodes every 3-4 years in "lemming booms"
- **Food Web:** Crucial prey for many Arctic predators
- **Tundra:** Help shape tundra vegetation through their grazing
- **Vocalizations:** Make various sounds including chirps, squeaks, and even barks
- **Food:** Can eat up to their body weight in plants daily
- **Myth:** Despite popular myths, they don't commit mass suicide by jumping off cliffs
- **Cycles:** Their boom-bust cycles significantly impact the entire Arctic ecosystem

Spotlight: Arctic Tern

Meet the Arctic tern, a small seabird with an incredible superpower: it makes the longest migration of any animal on Earth! Every year, these remarkable birds fly from the Arctic to the Antarctic and back again, covering a distance of up to 44,000 miles (70,800 km). That's like flying around the world twice! These long-distance champions spend their lives chasing the summer, experiencing more daylight than any other creature on the planet.

About Arctic Terns:

- **Scientific name:** *Sterna paradisaea*
- **Size:** 13-15 inches (33-39 cm) long
- **Wingspan:** 26-30 inches (66-77 cm)
- **Weight:** 3.2-4.2 ounces (90-120 g)
- **Lifespan:** Up to 30 years
- **Diet:** Mainly small fish and marine invertebrates
- **Habitat:** Arctic tundra in summer, Antarctic waters in winter
- **Migration:** Up to 44,000 miles (70,800 km) annually
- **Nest defense:** Aggressive, diving at intruders
- **Mating habits:** Mate for life, return to same breeding grounds
- **Flight adaptations:** Forked tail and pointed wings for long flights
- **Sleep habits:** Can sleep while flying, resting half their brain at a time
- **Hunting ability:** Excellent vision to spot fish from high altitudes
- **Vocalization:** Variety of calls, including distinctive "kee-arr" sound
- **Nesting:** On the ground in large colonies
- **Chick development:** Can fly within a month of hatching
- **Conservation status:** Threatened by climate change impacts

DID YOU KNOW?
An Arctic tern can fly more than 1.5 million miles (2.4 million km) during its lifetime – that's like traveling to the Moon and back three times!

Spotlight: Snow Goose

The snow goose is a magnificent bird that paints the sky white during its massive migrations. These large, striking birds breed in the Arctic tundra and then fly south in huge flocks that can number in the hundreds of thousands. Their annual journeys between the far north and southern winter grounds are a breathtaking spectacle of nature, filling the air with a cacophony of honks and the sight of countless white-feathered bodies against the blue sky.

About Snow Geese:

- **Scientific name:** *Anser caerulescens*
- **Size**: 25-31 inches (63-79 cm) long
- **Wingspan:** 53-65 inches (135-165 cm)
- **Weight:** 4.5-6.6 pounds (2-3 kg)
- **Lifespan:** Up to 20 years in the wild
- **Diet:** Mainly plant matter, including grasses, sedges, and grains
- **Habitat:** Arctic tundra in summer, marshes and agricultural fields in winter
- **Migration:** Travel up to 3,000 miles (4,800 km) each way annually
- **Breeding:** Form monogamous pairs that can last for years
- **Nesting:** Build nests on the ground, often on slightly raised areas
- **Clutch size:** 2-6 eggs per nest
- **Goslings:** Can walk and swim shortly after hatching
- **Flight formation:** V-shaped flocks for energy efficiency
- **Population trend:** Numbers have increased dramatically in recent decades
- **Conservation impact:** Large populations can damage Arctic breeding habitats
- **Vocalization:** Distinctive honking calls, especially during migration

DID YOU KNOW?

Snow geese have excellent eyesight and can see much farther than humans. This keen vision helps them spot predators and navigate during their long migrations across vast distances!

Spotlight: Harp Seal

Harp seals are marine mammals known for their striking coat patterns and their amazing ability to thrive in the icy waters of the Arctic and North Atlantic. These agile swimmers are famous for their white-coated pups, often called "whitecoats," which blend perfectly with the snow and ice where they're born. Harp seals play a crucial role in the Arctic ecosystem and face unique challenges as climate change affects their icy habitats.

About Harp Seals:

- **Scientific name:** *Pagophilus groenlandicus*
- **Size:** 5-6.5 feet (1.5-2 m) long
- **Weight:** 260-330 pounds (118-150 kg), males slightly larger
- **Lifespan:** Up to 35 years in the wild
- **Diet:** Mainly fish and crustaceans
- **Habitat:** Arctic and sub-Arctic waters of the North Atlantic and Arctic Oceans
- **Diving ability:** Can dive up to 1,000 feet (300 m) deep
- **Pup coat:** Born with famous white coat, which changes color after 3-4 weeks
- **Molting:** Adults shed and regrow their coat each year
- **Migration:** Travel long distances between breeding and feeding grounds
- **Breeding:** Form large colonies on pack ice to give birth and mate
- **Predators:** Polar bears, Greenland sharks, and occasionally orcas
- **Adaptation:** Thick blubber layer for insulation in cold waters
- **Social behavior:** Gather in large groups during breeding and molting seasons
- **Vocalization:** Use various underwater calls to communicate
- **Conservation status:** Least Concern, but vulnerable to climate change
- **Identification:** Adults have distinctive harp-shaped markings on their backs

DID YOU KNOW?

Harp seals have excellent underwater vision and can close their nostrils when diving.

Fun with Mazes

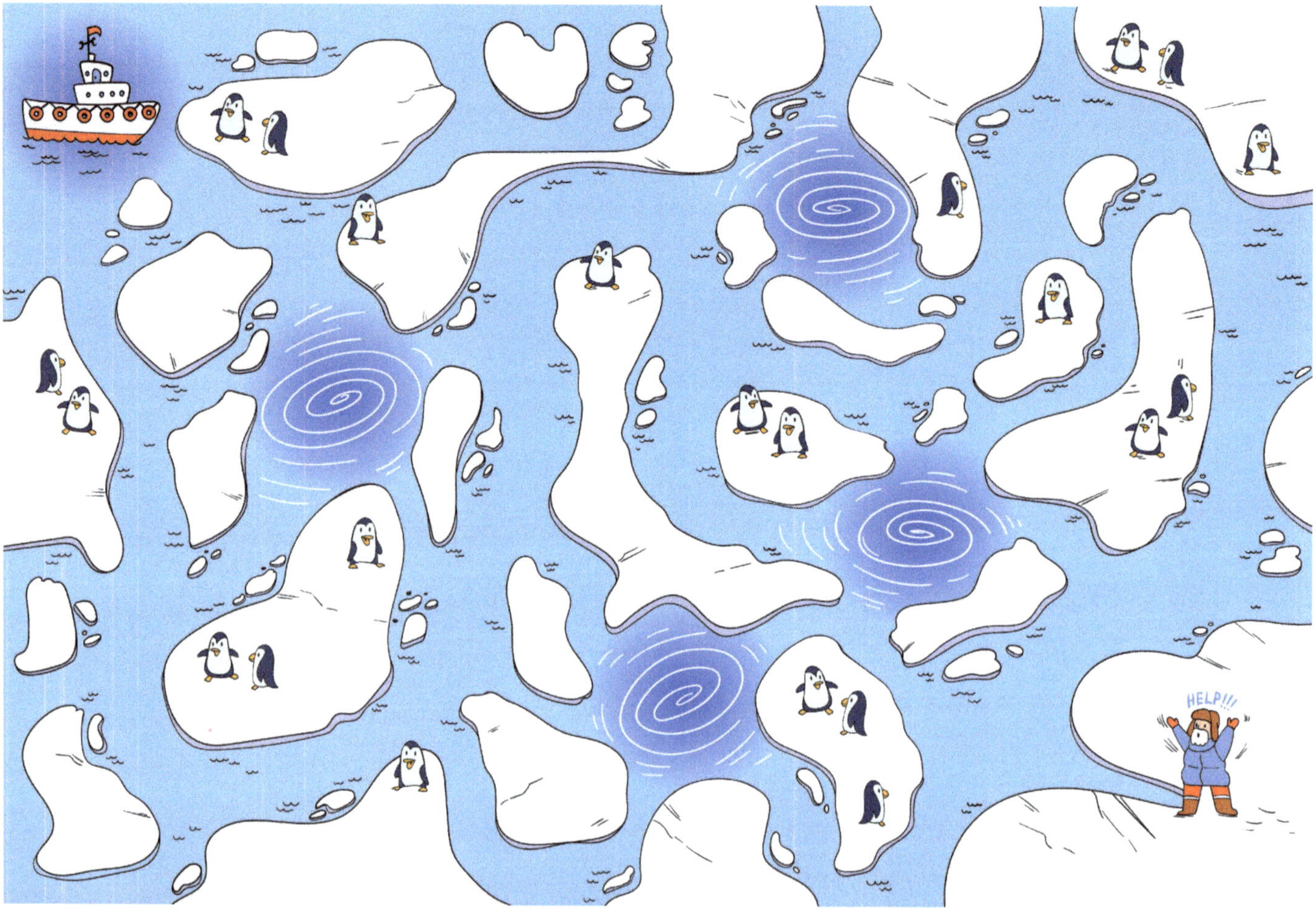

Crossword Challenge

Across

5 This Arctic region, covered in permafrost, is a cold, treeless plain where only hardy plants can grow (6)

7 This clever Arctic mammal changes its fur color with the seasons for perfect camouflage. (6,3)

10 This thick layer of fat beneath the skin of some Arctic animals (7)

11 This special layer of soil, found in the Arctic, remains frozen year-round (10)

13 Large, floating sheets of ice are a common sight in the Arctic Ocean (8)

14 This period of continuous daylight occurs in the Arctic during the summer months (8,3)

15 White-furred hunters, known for their keen senses and pack behavior (6,6)

Down

1 A quick hopper with large feet and white fur (6,4)

2 Known as the "unicorn of the sea" (7)

3 These Arctic giants use their long tusks to haul themselves out of the water and onto the ice. (8)

4 Top predator of the Arctic (5,4)

6 Natural light display, often seen in the Arctic night sky, is also known as the Aurora Borealis (8,6)

8 Arctic foxes use this to stay hidden from predators and sneak up on prey, changing with the seasons (10)

9 These vocal whales, often called the "canaries of the sea," are known for their varied sounds (7)

11 Colorful-beaked seabird that "flies" underwater in search of fish (6)

12 Known for its epic migrations and is called caribou in North America (8)

Find the Hidden Words

J I I A D A P T A T I O N S T L N N
X R I N G E D S E A L O Q N S O I T
F I B Q D B T S O R F A M R E P A I
N U F M V E W O L V E S E R O N W R
H B E R C L K J M J E I P T H U V D
W K Q N N U K M B H C E Z Y X S N P
A V O J O G P X U A A F Z E X T C D
L E L T I A Q O L S L R G S Y H N Z
R F L R T W A G L N K A E T I G F W
U J Z E A H O R S A L O I S S I N B
S T Y B R A D G C F R S X E G N A I
S W R B G L H H U T R B X E Z D R Y
N J O U I E V O G E I O E R N I W T
O S T L M S M R V O F C R A Y M H X
W A I B L A P I F Z Q E O O R D A H
Y L R M C V D I F C C K U C L S L I
O F R Q R O B E C O S Y S T E M S A
W Z E Z I D T H U O B I R A C A Y B
L O T B Y O S N I F F U P R Z J N F
S K I Q E G N A H C E T A M I L C S

Adaptations
Arctic Ocean
Beluga whale
Biodiversity
Blubber
Camouflage
Caribou
Climate change

Ecosystem
Foxes
Glaciers
Hares
Midnight sun
Migration
Musk oxen
Narwhals

Permafrost
Polar bears
Puffins
Ringed seal
Snowy owls
Territory
Walrus
Wolves

Find the 10 Differences

Glossary

Adaptations: Special characteristics that help animals survive in their environment, like thick fur for Arctic animals.

Alpha: The leader of a pack of animals, such as wolves in the Arctic.

Biodiversity: The variety of living things in a particular place, like the many dif ferent animals in the Arctic.

Bradycardia: When an animal's heart beats very slowly, which helps some Arctic animals save energy during the winter.

Camouflage: When an animal's color or pattern helps it blend in with its surroundings, like an Arctic hare's white fur in the snow.

Diurnal: Animals that are active during the day, like some Arctic birds.

Echolocation: A way some animals, like beluga whales, find their way around and locate food using sound.

Ecosystem: A community of living things and their environment, like the Arctic tundra.

Hemoglobin: A protein in the blood that carries oxygen. Some Arctic animals have higher con centrations, allowing them to dive longer and deeper.

Herbivores: Animals that eat only plants, like Arctic hares and lemmings.

Incubation: The time when a parent bird sits on its eggs to keep them warm until they hatch.

Irruption: When a large number of animals suddenly move to a new area, often in search of food, like snowy owls sometimes do.

Keystone species: An animal that plays a very important role in its ecosystem, like the ringed seal.

Lanugo: Soft, downy fur that some baby Arctic animals are born with to keep them warm.

Migration: When animals travel to a different place, usually to find food or have their babies.

Navigation: The way animals find their way from one place to another, often using the sun, stars, or Earth's magnetic field.

Omnivores: Animals that eat both plants and other animals.

Permafrost: Ground that stays frozen all year round, which is common in the Arctic.

Plumage: The feathers of a bird, which can help Arctic birds stay warm and recognize each other.

Precocial: Baby animals that are born well-developed and can move around and find food on their own, like Arctic hares.

Submerged: When something is underwater, like some Arctic plants that grow beneath the ice.

Territory: An area that an animal, like an Arctic wolf, defends as its own.

Vocalizations: The sounds animals make to communicate with each other, like the songs of Arctic whales.

Index

Sources and Additional Resources

Arctic Council. "Biodiversity." https://arctic-council.org/en/explore/topics/biodiversity/.

Blix, Arnoldus Schytte. "Arctic Animals and Their Adaptations to Life on the Edge." Berlin: Springer, 2005.

Canadian Geographic. "Arctic Species." https://canadiangeographic.ca/topics/arctic-species/.

Chester, Sharon. "The Arctic Guide: Wildlife of the Far North." Princeton: Princeton University Press, 2016.

International Union for Conservation of Nature (IUCN). "The IUCN Red List of Threatened Species." https://www.iucnredlist.org/.

National Geographic. "Arctic Animals." https://www.nationalgeographic.com/animals/topic/arctic-animals.

National Oceanic and Atmospheric Administration (NOAA). "Arctic Theme Page." https://www.arctic.noaa.gov/.

Polar Bears International. https://polarbearsinternational.org/.

Polar Research Board, National Academies of Sciences, Engineering, and Medicine. https://www.nationalacademies.org/prb/polar-research-board.

Sale, Richard. "A Complete Guide to Arctic Wildlife." London: Christopher Helm, 2006.

Sibley, David Allen. "Arctic Birds of North America: The Complete Guide." New York: Knopf, 2016.

The Arctic Institute. https://www.thearcticinstitute.org/.

World Wildlife Fund (WWF). "Arctic Wildlife." https://www.worldwildlife.org/places/arctic.

Activity Answers

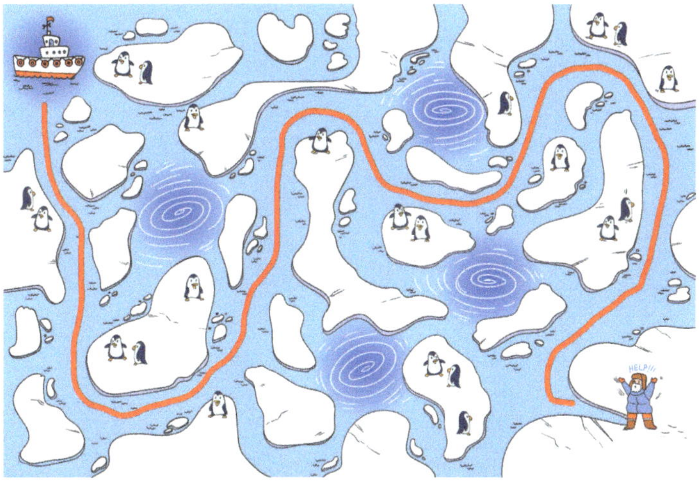

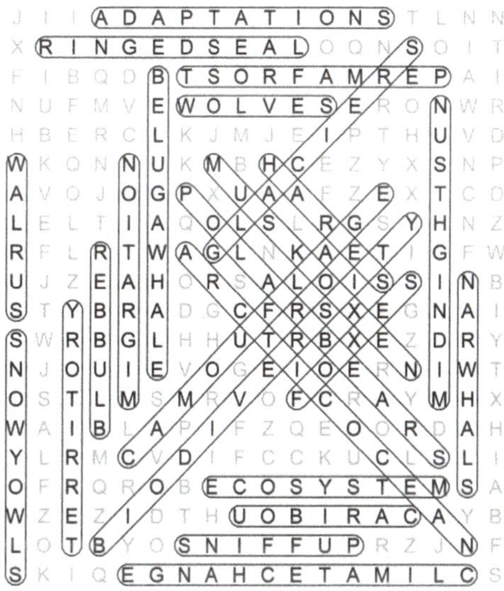

www.ingramcontent.com/pod-product-compliance
Ingram Content Group UK Ltd.
Pitfield, Milton Keynes, MK11 3LW, UK
UKHW060120010826
14089UKWH00005B/83

9 781647 904807